INTRODUCTORY COLLEGE MATHEMATICS

Tables and Graphs

ROBERT D. HACKWORTH, Ed.D.
Department of Mathematics
St. Petersburg Junior College at Clearwater
Clearwater, Florida

and

JOSEPH HOWLAND, M.A.T.
Department of Mathematics
St. Petersburg Junior College at Clearwater
Clearwater, Florida

SAUNDERS
SERIES IN **M**ODULAR
MATHEMATICS

W. B. Saunders Company: West Washington Square
                          Philadelphia, PA 19105

                          12 Dyott Street
                          London, WC1A 1DB

                          833 Oxford Street
                          Toronto, Ontario M8Z 5T9, Canada

INTRODUCTORY COLLEGE MATHEMATICS　　　　　ISBN 0-7216-4421-X
Tables and Graphs

©1976 by W. B. Saunders Company. Copyright under the International Copyright Union. All rights reserved. This book is protected by copyright. No part of it may be reproduced, stored in a retrieval system, or transmitted in any form or by any means, electronic, mechanical, photocopying, recording, or otherwise, without written permission from the publisher. Made in the United States of America. Press of W. B. Saunders Company. Library of Congress catalog card number 75-23628.

Last digit is the print number:　9　8　7　6　5　4　3　2　1

PREFACE

Tables and Graphs

This book is one of the sixteen content modules in the Saunders Series of Modular Mathematics. The modules can be divided into three levels, the first of which requires only a working knowledge of arithmetic. The second level needs some elementary skills of algebra and the third level, knowledge comparable to the first two levels. *Tables and Graphs* is in level 1. The groupings according to difficulty are shown below.

Level 1	Level 2	Level 3
Tables and Graphs	*Numeration*	*Real Number System*
Consumer Mathematics	*Metric Measure*	*History of Real Numbers*
Algebra 1	*Probability*	*Indirect Measurement*
Sets and Logic	*Geometric Measures*	*Algebra 2*
Geometry	*Statistics*	*Computers*
		Linear Programming

The modules have been class tested in a variety of situations: large and small discussion groups, lecture classes, and in individualized study programs. The emphasis of all modules is upon ideas and concepts.

Tables and Graphs is appropriate for all non-science majors especially social science students. The module is also well suited for math-science and technical students.

Tables and Graphs begins by explaining the common properties of tables and graphs. The emphasis is on skill in reading and constructing bar graphs, line graphs, circle graphs, and scattergrams. Also included is the use of graphing in selecting optimum opportunities in situations with a number of alternatives.

In preparing each module we have been greatly aided by the valuable suggestions of the following excellent reviewers: William Andrews, Triton College, Ken Goldstein, Miami-Dade Community College, Don Hostetler, Mesa Community College, Karl Klee, Queensboro Community College, Pamela Matthews, Chabot College, Robert Nowlan, Southern Connecticut State College, Ken Seydel, Skyline College, Ara Sullenberger, Tarrant County Junior College, and Ruth Wing, Palm Beach Junior College. We thank them and the staff at W. B. Saunders Company for their support.

Robert D. Hackworth
Joseph W. Howland

NOTE TO THE STUDENT

OBJECTIVES:

Upon completing this unit, the reader is expected to be able to demonstrate the following concepts and skills:

1. An ability to read and compare tabular data.

2. An ability to read and construct bar graphs, line graphs, and circle graphs.

3. An ability to use and construct problem solving graphs of business situations.

4. To evidence knowledge of trend lines and the correlation involved with scattergrams.

5. An ability to find the value of money invested at compound interest.

Three types of problem sets, with answers, are included in this module. Progress Tests are always short with four to six problems. The questions asked in Progress Tests come directly from the material of the section immediately preceding the test.

Exercise Sets appear less frequently in the module. More problems are in an Exercise Set than in a Progress Test. These problems arise from all sections of the module preceding the Exercise Set. Problems in Section I of each Exercise Set are specifically chosen to meet the objectives of the module. Problems in Section II of each Exercise Set are challenge problems.

A Self-Test is found at the end of the module. Self-Tests contain problems representative of the entire module.

In learning the material, the student is encouraged to try each problem set as it is encountered, check all answers, and restudy those sections where difficulties are discovered. This procedure is guaranteed to be both efficient and effective.

CONTENTS

Introduction..1

The Basic Concept of a Table.................................2

Bar Graphs...6

Line Graphs...14

Problem-Solving Graphs......................................18

Scattergrams..26

Circle Graphs and Pictographs...............................31

Percent and Interest Tables.................................41

Module Self-Test..47

Progress Test Answers.......................................53

Exercise Set Answers..55

Module Self-Test Answers....................................61

TABLES AND GRAPHS

INTRODUCTION

With the possible exception of basic arithmetic, no topic in mathematics is as important for the average citizen as an ability to read and interpret tables and graphs. The daily newspaper contains a variety of tables and graphs. Almost any non-fictional book uses tables and graphs. Even the visual displays on television frequently contain graphs and tables to better communicate their messages.

There are three major reasons for the importance of tables and graphs. First, materials in tables and graphs are organized and classified in a manner designed to promote understanding. Organizing and classifying data is a major step in any problem-solving or analysis situation. To the extent that a table or graph provides appropriate groupings to its data, the table or graph enables the reader to achieve a faster, easier understanding of its material.

A second reason for the importance of tables and graphs is the visual appeal they have for the reader. An old newspaper cliche is "a picture is worth a thousand words" and that quote is true in reporting numerical data as it is for any other newspaper subject.

The third reason for the importance of tables and graphs is the extent to which they are used in the evaluation, prediction, and decision-making processes. It is commonly believed that tables and graphs simply transmit information, but that view is unnecessarily narrow. In this module, an effort is made to broaden the uses or tables and graphs and show their value for disclosing similarities, contrasts, and future trends.

The content of the tables and graphs of this module came from current events with special efforts made to choose topics which are of broad general interest. Topics include the environment, the economy, the costs of a college education, the physical growth of adolescents, and future population projections. The tables in the latter half of this module show both business applications and computational tools for the mathematician.

THE BASIC CONCEPT OF A TABLE

The basic idea of most graphs and tables is to show a relationship between two or more categories of information. This is often accomplished by representing one type of information in a vertically written list and the other type of information in a horizontally written list. Each position in a table shows the relationship between its horizontal and vertical headings. The information or numbers shown in a table are called entries.

Table 1 illustrates the idea of horizontally and vertically listed categories of information. The table shows federally funded programs for the improvement of the environment and public health. The first item in the vertical listing is "Improved Public Transportation." The last item in the vertical listing is "Research on Pollutants." The first item in the horizontal listing is "Fiscal Year 1972" and the last item is "Percent Change."

TABLE 1

FEDERALLY FUNDED PROGRAMS IN DOMESTIC
HEALTH AND IMPROVEMENT OF THE ENVIRONMENT

(All amounts in millions of dollars)

	Fiscal Year 1972	Fiscal Year 1973	Change 72-73	Percent Change
Improved Public Transportation	$456	$666	$+210	46
Cancer Research	337	430	+93	27
Production of Electricity without Pollution	392	480	+88	22
Research & Development in Education	142	197	+55	39
Safety Programs for Natural Disasters	93	136	+43	46
Poverty Programs	103	141	+38	31
Research on Pollutants	115	154	+39	34

The table of federal programs correlates some of the methods for improving the health and environment with the amounts of money provided for the separate programs. According to the table, 337 million dollars were spent in fiscal year 1972 for cancer research. This information is found using the horizontal row of "Cancer Research" and the vertical column of "Fiscal Year 1972." Where the row meets the column, is the entry "337" which means 337 million dollars. Notice that directly below the title of the table there is a statement that each amount is in millions of dollars.

The table shows that the fastest growing program in terms of money spent was "Improved Public Transportation." This fact is established by comparing the numbers in the column headed "Change 1972-73." In terms of percent, the fastest growing programs are "Improved Public Transportation" and "Safety Programs for Natural Disasters" because each shows 46 percent change.

The next table is taken from an article comparing the economics of the southern United States with the United States as a whole. This table is representative of the presentation of data in many newspapers, magazines, and textbooks. As in the last example, the table is read using its horizontal listings in conjunction with the vertical listings. There are four columns in the table; two columns contain information for 1960 and the other two columns contain information for 1970. There are also two sections of horizontal information in the table. The upper section shows amounts of money in billions of dollars. The lower section shows the percent of distribution for the dollar amounts of the upper section.

PERSONAL INCOME IN THE UNITED STATES AND THE SOUTH, 1960 and 1970

	1960		1970	
	United States	South	United States	South
	(Billions)			
Farm	$ 15.0	$ 3.7	$ 19.0	$ 4.9
Nonfarm wage and salary	266.1	40.7	533.5	93.5
Other Labor income	11.0	1.6	30.8	5.1
Nonfarm Proprietor's income	34.2	5.6	51.0	8.7
Property Income	52.4	7.2	113.0	18.0
Transfer Payments	29.5	5.1	79.6	14.9
Less: Personal payments for Social Insurance	-9.2	-1.4	-28.0	-4.7
	$399.0	$62.3	$796.9	$140.4
	(Percent Distribution)			
Farm	3.8	5.9	2.4	3.5
Nonfarm wage and salary	66.7	65.2	66.8	66.6
Other Labor income	2.8	2.5	3.9	3.6
Nonfarm Proprietor's income	8.6	9.0	6.4	6.2
Property Income	13.1	11.5	14.1	12.8
Transfer Payments	7.4	8.2	10.0	10.6
Less: Personal payments for Social Insurance	-2.3	-2.3	-3.5	-3.4
TOTAL	100.00	100.00	100.00	100.00

Note: Detail may not add to total because of rounding.

Source: U.S. Department of Commerce, Office of Business Economics, Survey of Current Business, August, 1963 and August, 1971

The first horizontal line of the table shows that farm income in the United States in 1960 was 15 billion dollars, in the South in 1960 it was 3.7 billion dollars, in the United States in 1970 farm income was 19 billion dollars, and in the South in 1970, it was 4.9 billion dollars.

Property income is shown in the fifth horizontal line of the upper section of the table. That line shows that property income in the United States in 1970 was 113 billion dollars. Transfer payments were the sixth horizontal lines in both the upper and lower sections of the table. To find the percent of income attributable to transfer payments in the South in 1970, the reader should find the transfer payments line in the lower section of the table and read the entry in the last column on the right of the table. That entry is 10.6 and means that 10.6 percent of the personal income in the South in 1970 had been from transfer payments.

Progress Test 1

Below is shown a tabulated table in an advertisement. Use the table to answer the following questions.

1. What is the price of the machine that handles conversions?

2. How many models have the square roots feature?

3. If a customer wants a model only for adding, subtracting, multiplying, and dividing, which model is the most economical?

4. What two features are available on every model?

Features	Model No.	12R	21R	51R	61R	63R	80R	82R
	Price	$29.95	$49.95	$99.95	$79.95	$99.95	$139.95	$169.95
Adds, subtracts, multiplies, divides		✓	✓	✓	✓	✓	✓	✓
Algebraic logic			✓	✓	✓	✓		
Floating decimal		✓	✓	✓	✓	✓	✓	✓
Built-in rechargeability			✓	✓	✓	✓		
Memory			✓	✓✓	✓	✓		✓✓
Percent key			✓					✓
Square roots		✓			✓	✓		
Log and trig functions					✓	✓		
Scientific notation and parentheses						✓		
Conversions—metric, fractions, etc.				✓				
Printed tape							✓	✓

6   Introductory College Mathematics

BAR GRAPHS

In a graph the horizontal and vertical listings are usually arranged along two perpendicular lines. These lines are often called axes. In figure 1 below, a pair of axes is shown. The horizontal axis is labeled by years and the vertical axis is labeled by dollars.

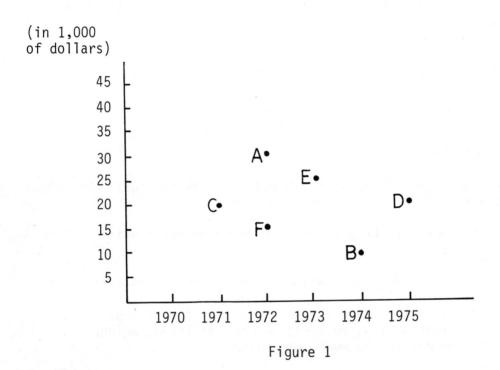

Figure 1

The point A on the graph represents two types of information. Because point A is directly above 1972 on the horizontal axis and directly across from 30 on the vertical axis, point A represents both 1969 and $30,000. (Note: numbers on the vertical axis represent thousands of dollars.) Point A therefore correlates or represents a relationship between the year 1972 and the dollar amount $30,000.

Similarly, point B in figure 1 represents both 1974 and $10,000. Each point of the graph shows a dual relationship between the years and the dollar amounts. Point B shows that for the year 1974 there is a matching dollar amount of $10,000 , but it also shows that for a dollar amount of $10,000 there is a matching year of 1974. The relationship is reversible because the year can be matched to the dollar amount or the dollar amount matched to the year.

Points C and D lie on the same horizontal line. Consequently, both points are associated with $20,000; C matches $20,000 to 1971 and D matches $20,000 to 1975. Notice that the points C and D match $20,000 to two different years, while the years 1971 and 1975 match to a single dollar amount.

Figure 2 is shown below and illustrates a simple bar graph.

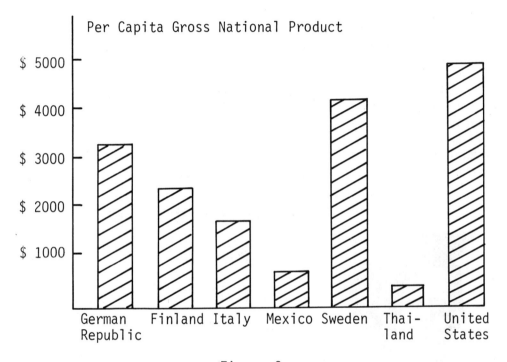

Figure 2

In Figure 2, the vertical axis is labeled in thousands of dollars and the horizontal axis lists seven selected countries. The heading for the graph is "Per Capita Gross National Product" which means the average economic production of the individual citizen.

The fourth vertical bar from the left represents Mexico. It is approximately two-thirds of the height of $1000 on the vertical axis and means that the per capita gross national product in Mexico is approximately $650.

8   Introductory College Mathematics

By comparing the heights of the bars in Figure 2, the greatest per capita income is found in the United States because it has the longest bar, and the smallest per capita income is found in Thailand because it has the shortest bar.

How does the average production of the citizens of the German Republic and Sweden compare? The average citizen in the German Republic produces about $3,300 and the average citizen in Sweden produces about $4,200 or $900 more per person. $900 is $\frac{900}{3,300}$ = 27% rounded to the nearest percent. The average citizen in Sweden produces about 27% more than the average citizen in the German Republic.

Figure 3 is another bar graph in which the horizontal axis is marked in years and each separate bar is labeled by a dollar amount. The graph shows the annual tuition costs in the public universities of Florida. Most of the bars in the table are darkened to indicate that figures for those years have been established. The bar for 1975-76 is indicated by an undarkened area with a dotted line. The meaning for the year 1975-76 is that this is a predicted amount. The table shows both data of the past and prediction for the near future.

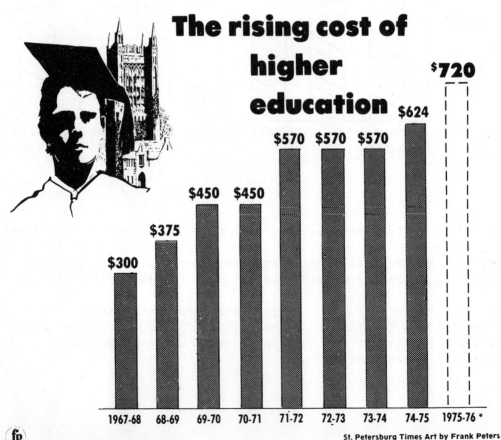

*Proposed, for juniors and seniors

Figure 3

St. Petersburg Times Art by Frank Peters

Progress Test 2

1. Use Figure 2 to rank in order, highest to lowest, the countries in terms of their per capita gross national products.

2. Use Figure 3 to find the year in which the greatest dollar increase in tuition costs was experienced.

3. Use the following information on per capita gross national product to make a bar graph like that in Figure 2.

    Canada $3676; France $2901; Israel $1836; Kenya $140;

    Japan $1911; Switzerland $3135; and Venezuela $979.

Exercise Set 1

I. 1. The table below shows the premiums on liability insurance for an accountant. The size of his total office staff affects the size of the premium.

ACCOUNTANT'S PROFESSIONAL LIABILITY
Rates for all states excepting New York

Total Staff	Deductible	$25,000	$50,000	$100,000	$250,000	$500,000	$1 Mill.
1	250.	80.	90.	100.	125.	160.	200.
2	250.	100.	112.	125.	156.	200.	250.
3	250.	120.	135.	150.	188.	240.	300
4	250.	140.	155.	175.	219.	280.	350.
5	250.	160.	180.	200.	250.	320.	400.
6	500.	—	194.	215.	269.	344.	430.
7	500.	—	207.	230.	288.	368.	460.

   a. What is the premium for $1,000,000 liability insurance for an accountant with a total office staff of 7?

   b. How much liability insurance can an accountant buy for $219 if there is a total of four people on his staff?

   c. Ms. Goldrock had $500,000 liability coverage when she had a total staff of four. After hiring three more people she raises the coverage to $1,000,000. How much more will she pay for the increased coverage?

10   Introductory College Mathematics

2. Use the table below to answer the questions following.

FAMILY INCOME DISTRIBUTION BY PERCENTAGE IN RACIAL MINORITY,
FOR THE NORTH AND WEST AND FOR THE SOUTH, 1959 AND 1969.*

	Percentage of Families with Low Incomes				Percentage of Families with High Incomes			
	1959 (below $3,000)		1969 (below $5,000)		1959 ($10,000 and over)		1969 ($15,000 and over)	
Percentage Minority	Whites	All Races	Non-blacks	All Races	Whites	All Races	Non-blacks	All Races
North and West								
0.0-9.9	11.9	12.7	12.3	12.9	19.3	18.0	23.7	23.1
10.0 and over	10.3	12.6	10.1	12.1	23.9	21.8	32.1	29.1
South								
0.0-24.9	18.7	23.1	15.7	19.1	15.6	13.6	21.1	18.7
25.0 and over	14.2	24.5	14.2	21.6	17.7	13.5	20.3	16.0

* 1959 units of analysis are Urbanized Areas; 1969 units are Standard Metropolitan Statistical Areas. Source of 1959 data: Glenn (1966), Table 2.

a. What percent of the white families in the North and West in the 0.0 - 9.9% minority group made less than $3,000 in 1959?

b. What percent of the families of all races made more than $10,000 is 1959 that were in the 25.0% and over minority in the South?

c. By how much percent did the all race family income percent change from 1959 to 1969 of the people that made high incomes in the North and West in the 10.0 percentage minority and over?

d. What percent of the non-black families in the 0.0 - 24.9 percent minority in the South made $15,000 or more in 1969?

e. Find the percent minority group and area that showed the greatest gain in percent of family income distribution in the all race, high income group from 1959 to 1969.

Tables and Graphs 11

3. Use the tire prices in the table below to answer the following questions:

SIZE	CURRENT SUGGESTED RETAIL PRICE		SPECIAL SALE PRICE		FEDERAL EXCISE TAX
	BLACK	WHITE	BLACK	WHITE	
5.60-15	$25.57	$28.88	$19.75	$23.88	$1.79
A-78-13	25.57	28.88	19.75	23.88	1.76
B-78-13	—	28.88	—	23.88	1.84
C-78-13	—	28.88	—	23.88	1.94
B-78-14	—	28.88	—	23.88	1.98
C-78-14	—	28.88	—	23.88	2.04
E-78-14	—	33.55	—	28.55	2.27
F-78-14	30.82	33.55	25.82	28.55	2.40
F-78-15	—	33.55	—	28.55	2.45
G-78-14	35.33	37.11	30.33	32.55	2.56
G-78-15	35.33	37.11	30.33	32.55	2.60
H-78-14	39.43	42.40	34.43	37.40	2.77
H-78-15	39.43	42.40	34.43	37.40	2.83
J-78-15	—	47.27	—	42.27	2.95
L-78-15	—	47.27	—	42.27	3.11

a. What is the current suggested price on an H-78-15 whitewall tire?

b. What is the sale price on a blackwall L-78-15 tire?

c. What is the difference between the sale price and the suggested retail price on a 5.60-15 blackwall tire?

d. What is the sale price of four whitewall E-78-14 tires including the Federal excise tax?

4. Use the bar graph showing the average of thirty industrial stocks to answer the following questions:

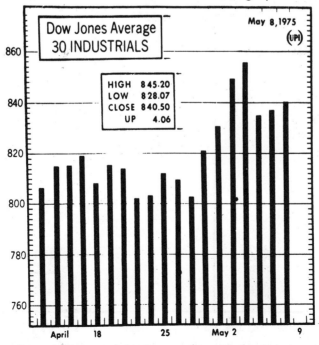

12   Introductory College Mathematics

    a. What was the Dow Jones Average on April 25, 1975?

    b. What was the trend of the average during the week just preceding May 2, 1975?

    c. How much did the Dow Jones Average increase from May 2 to the following market day?

5. Make a bar graph showing the following information:

   In 1970 the per capita gross national product was:

   $4,748 in the United States, $1,727 in Italy,

   $388 in Peru, $2,708 in Australia and $1,937 in Austria

6. Make a bar graph showing the following information on the number of people employed in agriculture given in thousands.

   1950 - 8,036   1955 - 6,945   1960 - 5,837   1965 - 5,128

   1970 - 4,932   1975 - 5,055

7. Use the bar graph on the deaths of active employees of the Metropolitan Life Insurance Company 1971-1972 to answer the questions on the opposite page.

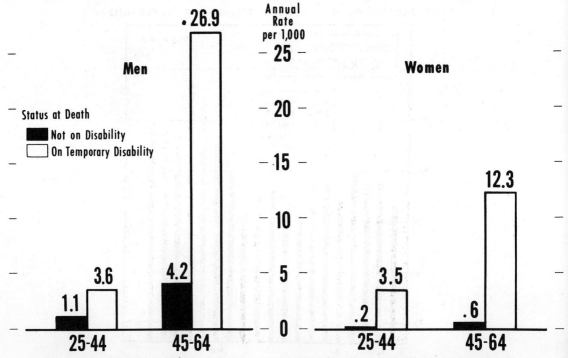

Deaths of Active*Employees-Metropolitan Life Insurance Company 1971-72

*Actively at work or on temporary disability. Note: Personnel in the Pacific Coast States and in Canada are not included.

a. What was the annual death rate for women 25-44 years old on temporary disability?

b. What was the annual death rate for men in the age group from 45 to 64 that were not on disability?

c. Which group ranks third in number of deaths in 1971-1972?

d. How does the death rate per thousand compare between the women age 45-64 not on disability and the men 45-64 not on disability?

II. Challenge Problems

The table on the right is an excerpt from the New York Stock Exchange Transactions as printed in the Wall Street Journal. Use the table to answer the questions.

—1975— High	Low	Stocks	Div	P-E Ratio	Sales in 100s	High	Low	Close	Net Chg.
10 5/8	7 1/8	BuddCo	.60	11	15	8 1/8	8	8	− 1/8
4 1/4	2 3/4	Budget	Ind	...	2	3	3	3	......
9 3/8	5 7/8	BulovaW	.20	4	22	6 1/4	6	6	......
22 1/8	18 1/8	BunkHil	1.86	...	15	18 3/4	18 3/8	18 5/8	− 1/8
8	3 1/4	BunkrR	.20p	...	79	5	4 3/4	4 3/4	− 1/4
28 1/4	14 7/8	Burlind	1.20	16	98	25 1/2	25 1/4	25 1/4	+ 1/4
42 7/8	29	BurlNo	.85p	10	175	31	29 1/2	31	+ 1 3/4
7 5/8	6 5/8	BurlNo	pf.55	...	14	6 7/8	6 5/8	6 5/8	− 1/8
32	11 3/4	**Burndy**	**.82**	10	5	29 1/2	29 1/2	29 1/2	......
110 3/4	61 1/8	**Burrghs**	**.60**	23	193	92 1/2	90 1/2	90 1/2	− 1 5/8

— C—C—C —

1. For BunkrR,

   a. what is its high for 1975?

   b. what was its high when the table was printed?

2. The P-E ratio of a stock is the ratio of the price of each share to the earnings for the year for each share. Using the closing price of Burroughs find its earnings per share for the year.

3. The earnings for each share of Burndy is what percent of the price?

4. If all the shares of Burlind sold the day of the newspaper printing sold at the closing price, what was the value of the shares sold?

5. If the earnings per share of a stock is 3% of the price of a share, what is the stock's P-E ratio?

14   Introductory College Mathematics

LINE GRAPHS

The line graphs of this section are introduced by another example of a bar graph. Figure 4 shows a different type of bar graph from the preceding section. In Figure 4 each bar represents the range of temperatures for its day. The tops of the seven bars shows the high temperatures for the days and the bottoms of the bars show the low temperatures.

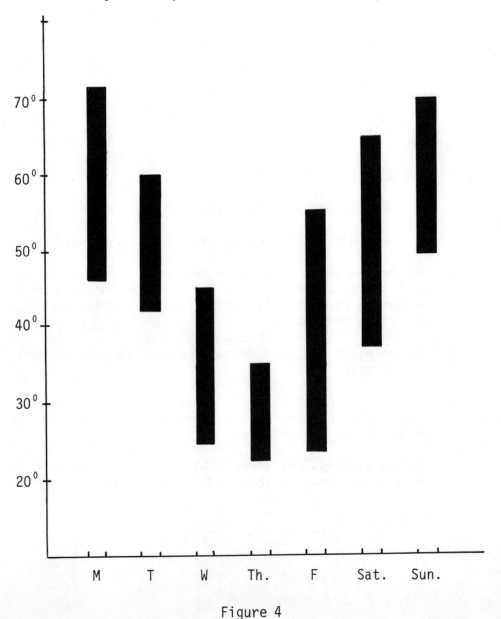

Range of temperatures over a one-week period.

Figure 4

If the midpoints of the seven bars of Figure 4 are joined by a line, the result is the line graph shown below in Figure 5.

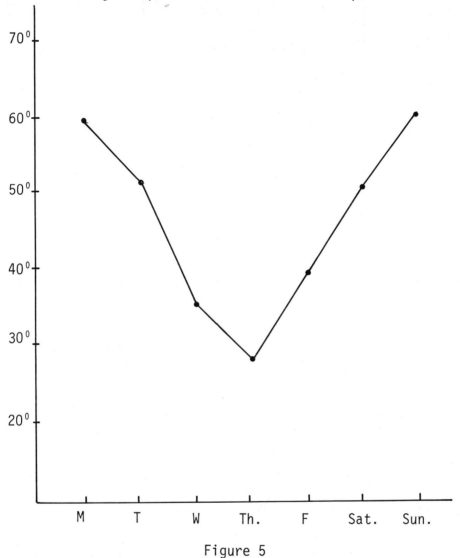

Figure 5

Notice that Figure 5 is a line graph of average temperatures whereas Figure 4 is a bar graph of the range of temperatures. In many instances the type of graph chosen for a particular set of data will depend on the purposes of the grapher. If average temperatures are desired, the line graph of Figure 5 is probably superior to the bar graph of Figure 4.

16   Introductory College Mathematics

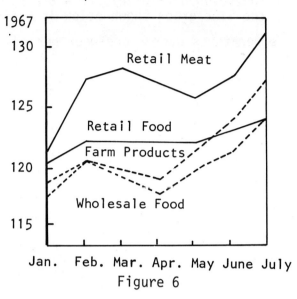

Figure 6

The vertical axis of Figure 6 is marked off in units of 5 index points where the 1967 prices of these food categories has been used to start the index at 100. This means that an index score of 120 represents 120% of the price in 1967.

The horizontal axis of the graph is marked in months and, by implication, the reader can assume there was one measurement per month selected as the index point for that month. Graphs like Figure 6 show the changes in each category of food and also provide the reader with a pictorial relationship between category prices.

Perhaps the most interesting overall information displayed by Figure 6 is the fairly close relationship between the prices of retail meat, farm products, and wholesale food. These three categories display some differences, but seem to generally rise and fall together. At the same time, retail food seems to maintain its own independent course with no decreases in the six-month period, but smaller overall increases than any of the other three categories.

The last line graph to be explained in this unit is shown in Figure 7. The vertical axis is marked in units of 50 million people. The horizontal axis is marked in units of one decade or ten years.

Tables and Graphs   17

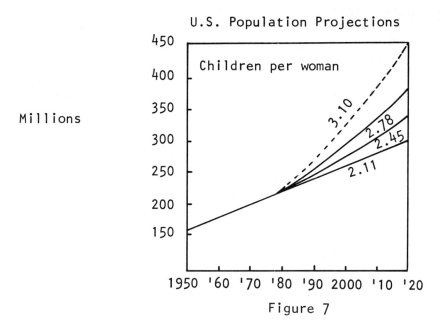

Figure 7

The interesting difference between the graph of Figure 7 and the earlier examples, is that Figure 7 is predicting future information while its predecessors were primarily reporting past data.

Figure 7 predicts the population of the United States through 2020 on the basis of four different rates of children per woman. For example, if the number of children per woman averages 2.78 and it is desired to find when the population will reach 350 million, then the solid line graph would be followed until it reaches the 350 level and the horizontal axis would then give the correct answer, 2010.

The reader interested in population control may know that the present rate of children per woman is between 2.11 and 2.45 which means that the projected U.S. population for 2020 is between 305 and 350 million people.

Progress Test 3

1. Use figure 6 to find:

    a. The difference between the index of retail meat for January and July.

    b. The difference in May of the index for Wholesale Food and Retail Food.

18   Introductory College Mathematics

2. Use Figure 7 to find:

   a. The projected time when U.S. population will surpass 300 million at a growth rate of 3.10 children per woman.

   b. The projected population of the U.S. in 1990 at a growth rate of 2.11 children per woman.

3. Use the bar graph of Figure 3 to construct a line graph showing the same information.

PROBLEM-SOLVING GRAPHS

The graph explained in this section is a composite of three line graphs. It differs from the earlier examples of line graphs because its use is to aid in a decision-making process.

A business situation is described below. The situation contains a problem for the manager of a manufacturing business and a graph will be used to solve the manager's problem.

THE BUSINESS SITUATION

The Handy Calculator Company manufactures and sells two types of hand calculators. Type A sells for $56. and Type B for $40. The company makes a maximum profit when it has a maximum income.

The manager knows of three limitations or constraints on his manufacturing and selling capabilities.

These restraints are:

1. The plant can manufacture a maximum of 900 calculators each week.

2. The sales force can sell no more than 700 Type B calculators each week.

3. The sales force can sell no more than 400 Type A calculators each week.

THE PROBLEM

The manager must decide how many Type A and Type B calculators should be manufactured to produce the greatest possible income.

The manager's decision would be easy if he could produce 1100 calculators each week. Since he can sell 700 Type B and 400 Type A calculators he would manufacture that many if his plant were not limited to a total production of 900 calculators.

The manager's decision would also be easy if he could sell all the calculators he could produce. Since Type A sells for more than Type B, an unlimited sales capacity would dictate manufacture of only Type A calculators. The fact that no more than 400 Type A calculators can be sold each week argues against producing more than that number of Type A calculators.

The manager is forced into a decision of producing both Type A and Type B calculators. His problem is to make the best choice of the numbers of Types A and B. That problem may be partially solved by the graph in Figure 8.

The horizontal axis is labeled in terms of Type A production possibilities. The vertical axis shows Type B possibilities. The three lines in the graph represent the limitations or constraints of the problem situation.

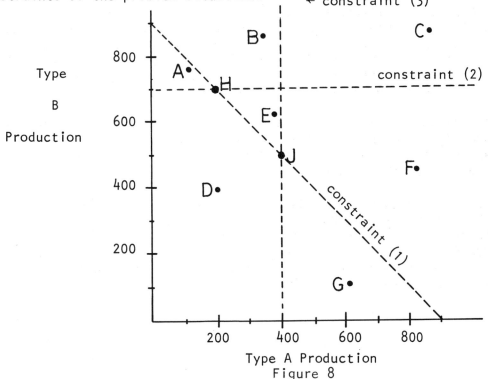

Figure 8

20   Introductory College Mathematics

1. Constraint (1) is a diagonal line representing the limitation of 900 as the maximum production for all calculators. Notice that the line of constraint (1) runs from 900 on the vertical axis to 900 on the horizontal axis.

2. Constraint (2) is the horizontal line intersecting 700 on the vertical axis. It represents the limitation of 700 on the sales of Type B calculators.

3. Constraint (3) is the vertical line intersecting 400 on the horizontal axis. It represents the limitation of 400 on the sales of Type A calculators.

The three constraint lines of the graph separate its area into seven sections represented by points with the capital letters A,B,C,D,E,F, and G.

Point A represents a poor managerial decision because it is above the line of constraint (2). Therefore, point A represents a production of more Type A calculators than the sales force can handle.

Point F also represents a poor management decision. Point F is above constraint (1) and to the right of constraint (3). Therefore, Point F represents a production of calculators over the 900 limitations and greater than the 400 Type A's that can be sold.

Because they violate at least one constraint, Points B, C, E, and G also represent poor management decisions. The only section of the graph that does not violate at least one constraint is in the section containing Point D. Point D itself is a poor management decision not because it violates any constraint -- but because greater production is possible.

The maximum income in this and all similar situations is represented by a corner (vertex) of the section that does not violate any constraints. Point H or Point J will represent the best choices for the manager.

Point H represents a production of 200 Type A calculators and 700 Type B calculators. The income from a management choice of Point H is found by determining the value of the calculators produced and sold. Each type A calculator is valued at $56; each Type B is valued at $40. The total value at Point H is figured on the opposite page

```
       200 Type A @ $56 is      $11,200

       700 Type B @ $40 is       28,000

       Total Value At Point H   $39,200
```

Point J represents a manufacture of 400 Type A and 500 Type B calculators. The income produced by Point J is figured below:

```
       400 Type A @ $56 is      $22,400

       500 Type B @ $40 is       20,000

       Total Value at Point J   $42,000
```

Comparing Point J income to Point H income solves the manager's problem. For maximum income the manager should manufacture and sell the calculators represented by Point J.

Progress Test 4

1. Use Figure 8 to determine the constraint violated by:

   a. Point B

   b. Point E

   c. Point G

2. Draw a decision-making graph for the following situation:

   A plant produces Type C and Type D calculators under the following constraints:

   1. A maximum of 120 calculators can be manufactured each week.

   2. Type C sales are limited to 90 or less each week.

   3. Type D sales are limited to 60 or less each week.

3. If Type C calculators are valued at $45 each and Type D at $52 each, find the rates of production for maximum income.

22 Introductory College Mathematics

Exercise Set 2

I. 1. Use the graph on raw steel production to answer the following questions.

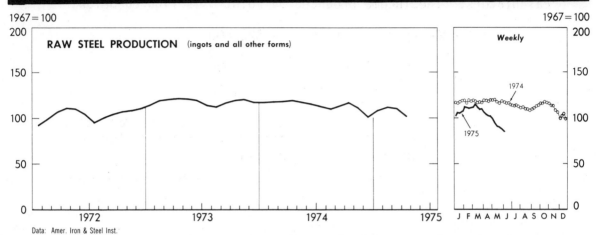

a. What is the general trend of raw steel production from 1972 to 1975?

b. What was the trend from February, 1975 to April, 1975?

c. The vertical scale shows the index number of steel production. What is the difference in the index numbers between April, 1974 and April, 1975?

d. What is the difference in the index numbers for weekly production between May, 1975 and May, 1974?

2. Use the graph on wholesale prices to answer the following questions:

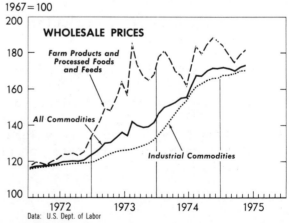

a. Which of the three groups had the most erratic price movement?

b. What was the difference between the prices of farm products and all commodities in January, 1975

c. What was the trend in the prices of industrial commodities from January, 1975 to April, 1975?

d. The prices of the three different groups were always different except for a time in the year _____.

3. Use the graph of manufacturing hours and earnings to answer the following questions:

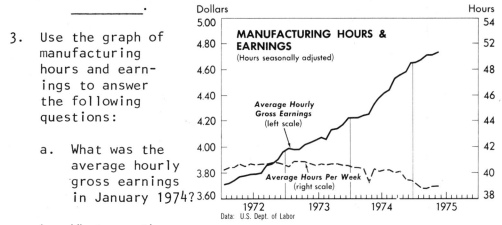

   a. What was the average hourly gross earnings in January 1974?

   b. What was the average number of hours worked per week in January, 1974?

   c. What is the comparison between the average gross earnings and the average hours worked per week during the years from 1972 to 1975?

4. Use the line graph of the spouse's mean scores on the Locke-Wallace Marital Adjustment Scale to answer the the following questions:

   a. What was the mean score for husbands on the 5th stage of the scale?

   b. Which stage of the adjustment scale had the greatest difference between the mean scores of the husbands and the wives?

   c. In general, how did the husbands, mean score on the scale compare with the wives, mean score?

   d. What was the wives, mean score on the postparental family stage of the scale?

   e. At which of the stages did the husbands score higher than the wives on the average?

24 Introductory College Mathematics

SPOUSES' MEAN SCORES ON THE LOCKE-WALLACE MARITAL ADJUSTMENT SCALE BY STAGE IN THE FAMILY LIFE CYCLE, NEWARK, OHIO

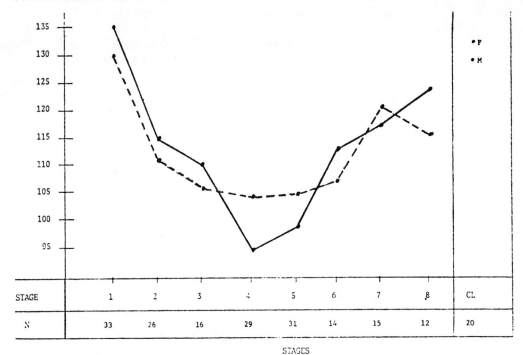

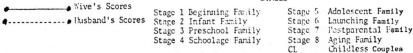

●────────● Wive's Scores
●--------● Husband's Scores

Stage 1 Beginning Family
Stage 2 Infant Family
Stage 3 Preschool Family
Stage 4 Schoolage Family
Stage 5 Adolescent Family
Stage 6 Launching Family
Stage 7 Postparental Family
Stage 8 Aging Family
CL Childless Couples

5. Use the following information to construct a modified bar graph similar to the range of temperature graph shown previously. The price of the following stock varied on a certain day as follows: National Utilities, 12 to 13, Ocean Oil, $27\frac{1}{2}$ to $38\frac{1}{2}$, Pinkerton, 35 to 40, Speidel, 10 to 18.

6. Make a line graph to show the mean scores of men and women on a series of tests given in freshman history.

	men	women
Test 1	88	92
Test 2	90	85
Test 3	85	86
Test 4	86	95

7. Use the following information to construct a line graph showing the number of colds people catch according to their age.

Age	1	1-2	3-4	5-9	10-14	15-19	20-24
Number of colds per year	6.1	5.7	4.9	3.4	2.6	2.3	2.8

8. A tire manufacturing company makes steel belted radial tires and glass belted bias ply tires at one of its plants. The radial tires sell for $70 apiece and the bias ply tires sell for $30 each. The plant has the following restrictions on its manufacturing and sales capabilities: (1) It can sell up to 1300 tires a week. (2) It can build up to 600 radial tires a week and (3) up to 1000 bias ply tires a week.

   a. Construct a graph showing the three restrictions on its capabilities.

   b. Which constraint is violated if it tries to sell 950 bias tires and 400 radial tires per week?

   c. Which restriction is violated if it tries to manufacture 650 radial tires and 900 bias tires per week?

   d. Find its gross income if it sells 1000 bias ply tires and no radial tires in a week.

   e. Find its gross income if it sells 700 bias tires and 600 radial tires a week.

   f. Find its income if it sells 1000 bias tires and 300 radial tires in a week.

   g. Find its gross income if it sells 600 radial tires and no bias tires in a week.

   h. How many radial tires and bias tires should the manager of the plant try to produce to maximize the plants income?

9. a. Make a decision-making graph to show the manufacturing and sales restriction of a tire plant that has the following constraints on its manufacture of nylon and polyester cord tires. It can sell no more than 20,000 tires a week. It can produce no more than 15,000 nylon cord tires a week and no

more than 18,000 polyester cord tires a week. The price of its nylon tires averages $22 and its polyester tires average $24 each.

b. Find its gross income if it sells 15,000 nylon and 5000 polyester tires a week.

c. If it sells 18,000 polyester tires in a week, how many nylon tires can it sell?

d. What is the maximum income that the plant can produce in a week and how many of each type tires must it sell to achieve the maximum income?

II. Challenge Problems

The Rent-a-Person Company has the following restrictions on its capabilities. It only rents out clerks and secretaries. It can rent out up to 80 clerks a day or up to 60 secretaries a day. The sum of twice the number of clerks and the number of secretaries is at most 170. It receives $50 a day for a clerk and $70 a day for a secretary. How many clerks and how many secretaries should it rent out to maximize its income and what is its maximum income?

SCATTERGRAMS

Frequently a list of people, businesses, sports teams, etc. will be accompanied by two characteristics for each element of the list and it is desired to show this information graphically. Such information can be shown with a scattergram.

For example, suppose ten men were chosen at random and each man's age and yearly income were found. There would be a list of ten men and each element of the list would have two characteristics: age and income. The results of such a survey might appear in the following scattergram.

Each X in Figure 9 represents one of the ten men surveyed.

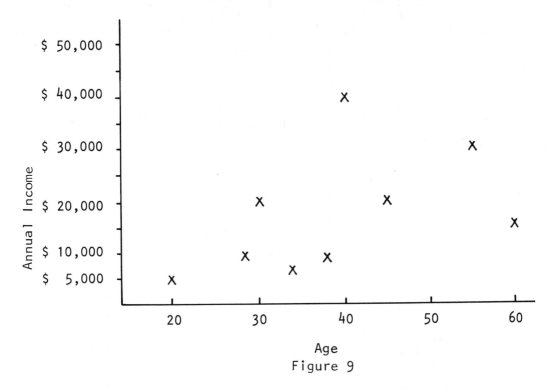

Age
Figure 9

The position of each X indicates both the age and annual income of the man. The man with the highest income is represented by the X directly across from $40,000 on the vertical axis. His age is approximately 40.

Each X in the scattergram represents two numbers -- an age and an annual income.

Figure 10 shows another scattergram taken directly from an educational psychology textbook. The scattergram shows two types of males: 16 early maturing boys are represented by triangles and 16 late maturing boys are represented by circles.

The scattergram of Figure 10 compares their heights at four different ages. The scattergram indicates that the physical effects of early or late maturing are limited to adolescence and do not carry over into adulthood. Late maturing males eventually attain approximately the same heights as their earlier growing counterparts.

28   Introductory College Mathematics

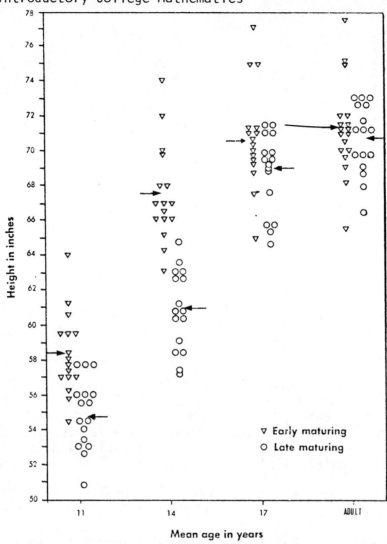

Figure 10

Figure 11 is a scattergram relating job satisfaction in a steel mill with the workers' distance from the blast furnace.

The scattergram of Figure 11 does not have X's spread completely over the graph. Rather the X's seem to be grouped along an oval shaped cluster from the lower left of the graph to the upper right. The further the worker is from the blast furnace, the higher his job satisfaction, and the higher the job satisfaction, the further his distance from the blast furnace. If a line is drawn in Figure 11 showing this relationship the result is Figure 12.

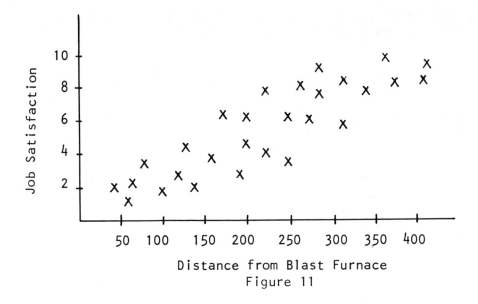

Distance from Blast Furnace
Figure 11

The line of Figure 12 is sometimes called a regression line. It shows the strong tendency for X's to group themselves close

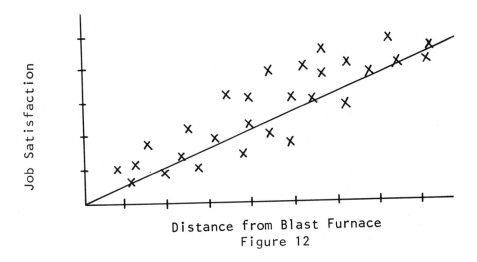

Distance from Blast Furnace
Figure 12

to the line. Whenever the items in a scattergram are grouped along a regression line, the two characteristics represented by the horizontal and vertical axes are said to correlate with each other. Perfect correlation is described by the number 1. Two characteristics would be in perfect correlation only when every item of the scattergram fell on the regression line.

Whenever the items of a scattergram do not cluster along a line but are spread out in the graph then the correlation between the characteristics is lessened. When the items are evenly spread across the entire space of the graph there is no correlation. The number used to show no correlation is the number 0.

Figure 13 shows four scattergrams and their related correlations represented by the small letter r. When r = 0.00 the items are spread evenly over the entire graph. As r increases to 0.90 the items tend to cluster more along an oblique line running from lower left to upper right.

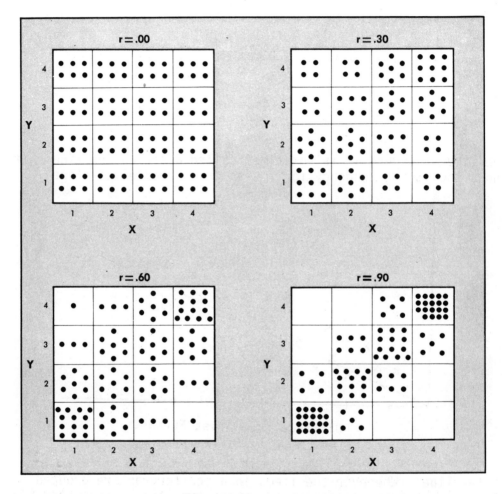

Figure 13

The degree of correlation between two characteristics can be of great value. When the correlation is high (close to 1) the knowledge of one characteristic is tantamount to knowing the value of the other characteristic. When the correlation is low (close to 0) the knowledge of one characteristic should not be used to predict a value for the other characteristics. The correlation between the two characteristics is a measure used in statistics and is quite difficult to compute.

Tables and Graphs   31

Progress Test 5

1. Draw a trend line (regression line) for Figure 9. Does there appear to be a correlation between the age and annual income of the men surveyed?

2. In Figure 10 some small arrows are shown. What do these small arrows indicate?

3. For what age does Figure 10 indicate there is a greatest height difference between early and late maturers?

4. Draw a scattergram using the following data (age-weight) for ten teenagers:

    (1) 16-145, (2) 18-150, (3) 17-105, (4) 15-112,

    (5) 16-125, (6) 14-140, (7) 15-130, (8) 17-130,

    (9) 15-124, (10) 16-165.

    a. Using X to represent the correlation between age and weight, does X tend to approach 1.00 or 0.00?

    b. What is the trend shown by the data?

CIRCLE GRAPHS AND PICTOGRAPHS

Circle graphs are usually used when fractional parts of a whole are to be compared. Figure 14 is a circle graph showing incomes of architects. Each portion or pie-shaped section of the circle represents a fraction of the architects. Since about one-fourth of the circle is labeled 15-20 the circle graph of Figure 14 indicates that approximately 25% of the architects earn between $15,000 and $20,000 yearly. The graph also shows that approximately the same fraction of architects earn between 8 and 12 thousand as those that earn more than $50,000.

32   Introductory College Mathematics

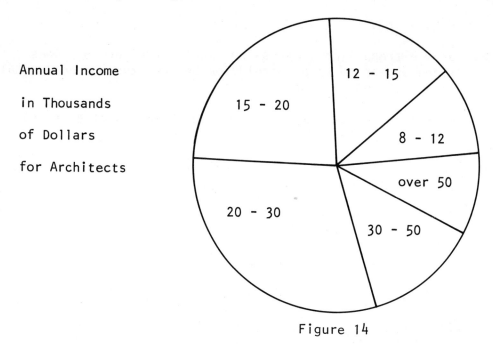

Figure 14

Circle graphs are extremely efficient at showing the relationship between the numbers that make up a sum. The numbers that are added to make a sum are called addends. Since each addend is a fraction of the sum, each portion of the associated circle graph represents a fraction of the total circle. Consequently, a circle graph shows graphically the comparison between the addends in a sum.

The construction of a circle graph depends upon knowledge of two facts:

1. There are $360°$ in a circle.

2. The percent of the sum that represents each addend of the sum.

If the percent representation of each addend is already known, the size of each portion of the circle graph can be found by multiplication. For example, a family decides to budget its income according to the following percents:

    Food   25%   Utilities   20%   Clothing 10%
    Housing   40%   Recreation   5%

The number of degrees that should be allotted to each portion of the budget can be found by multiplying the decimal equivalent of the percent by 360.

Food	0.25 x 360	=	90°
Housing	0.40 x 360	=	144°
Utilities	0.20 x 360	=	72°
Clothing	0.10 x 360	=	36°
Recreation	0.05 x 360	=	18°
Total			360°

Notice that food is 25% or $\frac{1}{4}$ of the total budget and 90° is also $\frac{1}{4}$ of the total circle of 360°. The reader can check the accuracy of his/her computations by totaling the degree column. If some of the addends in the degree column are rounded to the nearest degree, one of the items may need adjustment to ensure that the total is 360. None of the addends above needed rounding off, consequently the total was exactly 360°.

The circle graph illustrating the divisions of the previous budget can be constructed by first drawing a radius in a circle. Then the 90° angle representing the food portion of the budget can be measured from the first radius. The completed circle graph of the previous budget is shown in Figure 15.

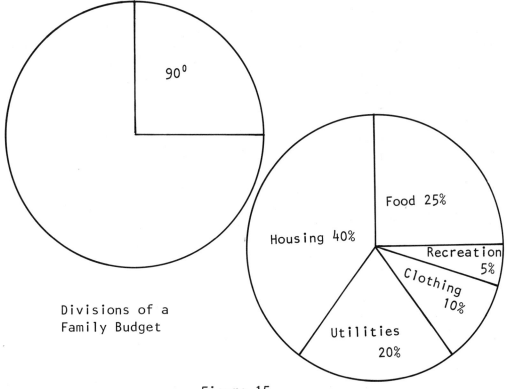

Divisions of a Family Budget

Figure 15

34   Introductory College Mathematics

In bar graphs the lengths of the bars are used to show the quantity of some characteristic. In pictographs small drawings or figures are used for the same purpose. The more small figures that are used the greater the characteristic.

In Figure 16 each figure of a girl represents 10% of the age group shown in the row. Three girl figures in the 13½ year old row therefore represent 30% of the girls at that age.

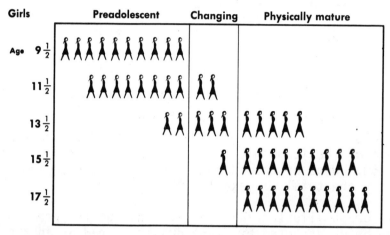

Figure 16

Figure 16 shows that 100% of the girls in the 9½ age bracket are in the preadolescent stage. By age 11½, 20% of the girls are changing while 80% remain in the preadolescent stage. Similarly, by age 13½ only 20% remain preadolescents, 30% are changing, and 50% have attained physical maturity.

Figure 17 is a pictograph map of the United States with three shadings to represent different approaches to branch banking laws. The darkest shading represents unlimited branch banking, the lighter shading represents limited branch banking, and the unshaded states allow no branch banking. The number of holding companies in each state appears above the percent.

Figure 17 shows that the far western states all allow unlimited branch banking. Kansas allows no branch banking. Wisconsin allows limited branch banking.

Tables and Graphs   35

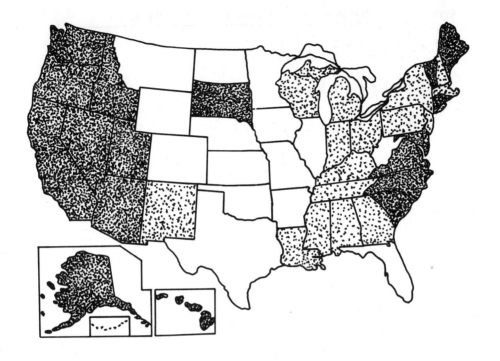

Unlimited Branching

Limited Branching

No Branching

Figure 17

Progress Test 6

1. Use Figure 16 and the graph on page 36 to answer:

    a. What percent of 15½ year old boys are physically mature?

    b. How do 13½ year old girls compare to 13½ year old boys in terms of their physical maturity?

    c. How do 17½ year old boys compare to 17½ year old girls in terms of their physical maturity?

36   Introductory College Mathematics

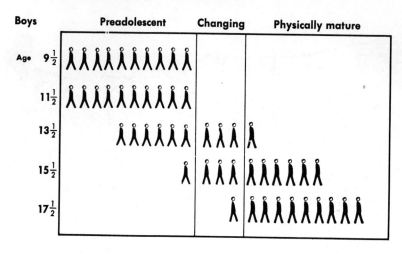

2. Use Figure 17 to answer:

   a. How many states allow no branch banking?

   b. What does South Dakota allow?

3. Construct a circle graph to show that for all contractors:

   a. 10% are over 60 years old

   b. 30% are 51-60

   c. 40% are 41-50

   d. 15% are 31-40

   e. 5% are 21-30

Exercise Set 3

1. Use the scatter gram on page 37 to answer the following questions:

   1. Batting Averages and Players' Ages

      a. Does there appear to be a correlation between the ages of the players and their batting averages? If so what is it?

      b. How many players are 26 years old?

      c. What is the best batting average in the 20 year old group?

Tables and Graphs   37

d. Projecting the trend of the scattergram, the batting average of a 32 year old player is likely to be _____ (higher than, lower than, same as) 250.

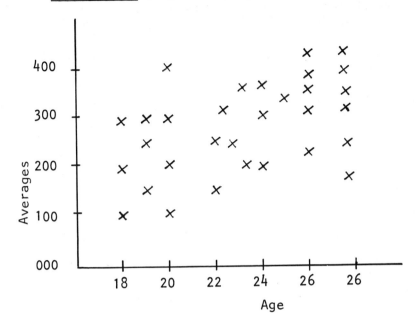

Sample Regression of Blood Pressure on Age. The broken lines indicate omission of the lower parts of the scales in order to clarify the relations in the parts occupied by the data.

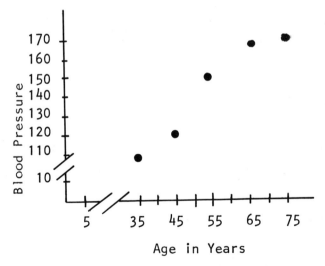

2. Use the graph above to answer the following questions:

   a. Which ten year period had the greatest rise in blood pressure?

38   Introductory College Mathematics

    b. Describe the trend of the regression line for the scattergram above.

    c. Which ten year period shows the least gain in blood pressure rise?

3. The four scattergrams below compare the age of a selected group of young people to their height, weight, number of relationships with the opposite sex and the number of their dentist visits. Use these graphs to answer the following questions:

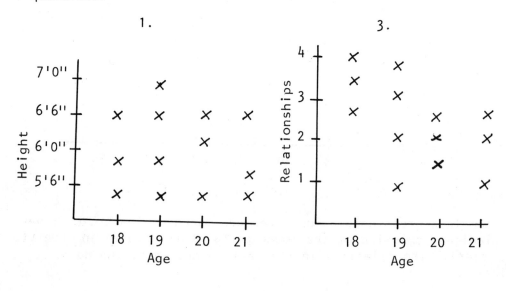

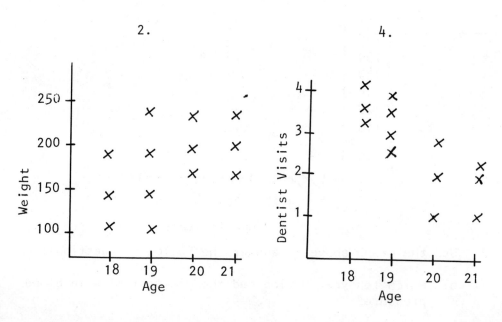

Tables and Graphs   39

   a. Which of the four graphs shows zero correlation?

   b. How do the regression lines for graph 3 and graph 4 compare?

   c. At what age does the greatest difference between the number of relationships with the opposite sex occur?

   d. Approximately how many dentist visits will an 18 year old person make?

   e. Which has the greatest correlation, age to dentist visits or age to weight?

   f. How much variation does there seem to be in a 20 year old person's weight?

4. Draw a scattergram, using the following data, on the age and self-confidence of a group of people:

   (1) 18 - 7   (5) 19 - 6   (9) 18 - 1
   (2) 20 - 7   (6) 20 - 4   (10) 21 - 5
   (3) 21 - 2   (7) 18 - 10   (11) 20 - 2
   (4) 19 - 4   (8) 21 - 9   (12) 19 - 8

   a. The correlation between age and self-confidence is _____ (high, low, about 0.5).

   b. Based on the scattergram above, what prediction could be made on the self-confidence of 22 year old people?

5. The graph below shows the percent of a certain population with a particular job satisfaction. Use the graph to answer the questions on page 40.

Each figure represents 12-1/2% of the population groups

40   Introductory College Mathematics

    a. How many people out of a group of 100 non-whites less than 18 years old like their job?

    b. What is the trend in job satisfaction of white people over their life time?

    c. What percent of the non-white people over 18 years old do not like their job?

    d. What percent of the non-whites under 19 change their attitude towards their job from positive to negative as they grow into the older age group?

    e. What is the trend in job satisfaction of the non-white people during their life time.

6. Make a circle graph showing the following information on the source of the budget dollar:

    Corporation income taxes - 24¢   Excise taxes - 9¢

    Individual income taxes - 39¢   Other - 11¢

    Employment taxes - 17¢

7. Make a circle graph showing the budget for the family below. The following allotments were made each month.

Housing	$300	Car Expense	$200
Clothing	$ 80	Savings	$175
Food	$250	Recreation	$ 45
Utilities	$150		

II. Challenge Problems

1. An 18 year old person has a 30 inch waist and a 20 year old person has a 32 inch waist. What will be the waist size of a 23 year old person if the correlation between waist size and age of people between 18 and 23 is 1.0?

2. A family allots $240 out of its monthly income of $1,700 for car expense. Find the size of the angle that should represent car expense for a circle graph of the family budget.

3. In a circle graph of the results of a political opinion survey, a 126° angle represented the 18 to 21 year old voters. There were 840 in the group of 18 to 21 year old voters. How many voters were surveyed?

PERCENT AND INTEREST TABLES

The final two sections of this module explain the uses of four mathematical tables. The two tables of this section might be used in any job situation where fractions, percent, or interest problems arise.

The table below is a fraction-decimal approximation chart. The table may be used to write many fractions as decimal approximations.

Fraction Numerators

Fraction Denominators	1	2	3	4	5	6	7	8	9	10	11	12	13	14	15
2	.50														
3	.33+	.67-													
4	.25	.50	.75												
5	.20	.40	.60	.80											
6	.17-	.33+	.50	.67-	.83+										
7	.14+	.29-	.43-	.57+	.71+	.86-									
8	.13-	.25	.38-	.50	.63-	.75	.88-								
9	.11+	.22+	.33+	.44+	.56-	.67-	.78-	.89-							
10	.10	.20	.30	.40	.50	.60	.70	.80	.90						
11	.09+	.18+	.27+	.36+	.45+	.55-	.64-	.73-	.82-	.91-					
12	.08+	.17-	.25	.33+	.42-	.50	.58+	.67-	.75	.83+	.92-				
13	.08-	.15+	.23+	.31-	.38+	.46+	.54-	.62-	.69+	.77-	.84+	.92+			
14	.07+	.14+	.21+	.29-	.36-	.43-	.50	.57+	.64+	.71+	.79-	.86-	.93-		
15	.07-	.13+	.20	.27-	.33+	.40	.47-	.53+	.60	.67-	.73+	.80	.87-	.93+	
16	.06+	.13-	.19-	.25	.31+	.38-	.44-	.50	.56+	.63-	.69-	.75	.81+	.88-	.94-

42   Introductory College Mathematics

The horizontal axis of the table is labeled along the top and shows the numerator or number above the bar of a fraction. The vertical axis shows the denominator or the number below the bar of a fraction.

To find the decimal approximation for $\frac{7}{9}$, the horizontal axis is used to find 7 and the vertical axis is used to find 9. The entry in the 7 column, 9 row is .78⁻. The minus sign behind .78⁻ means that $\frac{7}{9}$ is between .77 and .78 but is closest to .78. .78⁻ is a decimal approximation for $\frac{7}{9}$. The number .78⁻ is slightly less than .78

To find the decimal approximation for $\frac{3}{14}$ use the 3 column, 14 row. The entry is .21⁺. The plus sign behind .21⁺ means that $\frac{3}{14}$ is between .21 and .22 but is rounded off to .21. The number .21⁺ is a little larger than .21.

To find the decimal for $\frac{9}{12}$ use the 9 column, 12 row. The entry is .75. Since .75 is not followed by either a plus or a minus, the decimal .75 is equal to $\frac{9}{12}$.

As an example of the use for the table, consider the following situation:

> Smello Deodorant is sold in a 16 ounce aerosol can which contains 7 ounces of propellant. If the price is $1.35 per can, two questions of value received might be asked.
>
> 1. What is the price per ounce of the mixture?
>
> 2. What is the price per ounce of active ingredients (deodorant)?

To answer question 1. it is sufficient to find $\frac{1}{16}$ (16 ounces in the can) of $1.35. The table of fraction-decimal approximations can be used to find that $\frac{1}{16}$ is .06⁺. Multiplying 1.35 by .06 gives:

$$\begin{array}{r} 1.35 \\ \times .06 \\ \hline .0810 \end{array}$$

To answer question 2. it is sufficient to find $\frac{1}{9}$ (9 ounces of deodorant) of $1.35. The table has .11$^+$ as the decimal approximation for $\frac{1}{9}$. .11$^+$ is multiplied by 1.35.

$$\begin{array}{r} 1.35 \\ \times .11 \\ \hline 135 \\ 135 \\ \hline .1485 \end{array}$$

.1485 rounded off to $.15 or 15¢ is the cost per ounce of deodorant.

Percent means one-hundredths. Consequently, .63 is the decimal for 63%, .44$^+$ is slightly more than 44% and .17$^-$ is slightly less than 17%. The fraction-decimal table can be used to convert any of its fractions to percents. To write $\frac{5}{6}$ as a percent, find the decimal for $\frac{5}{6}$ and write it as a percent. $\frac{5}{6}$ = .83$^+$ and .83 = 83%

Many percent problems are involved in figuring the interest on a savings account, a credit card account, or a bank loan. The next table is a compound interest table that might be used in determining the value of a sum of money that accumulates at compound interest over a period of time.

Suppose, for example, that $10,000 is invested in a savings account at 5% annual interest to be compounded quarterly. "Compounded quarterly" means that the interest is computed four times per year and the amount of interest is added to investment. If the original investment were left in the savings account for 10 years, there would be forty (4 times 10) interest periods. To compute the value of the investment without a table or any mechanical assistance requires forty separate multiplications and many additions. The table is constructed to eliminate most of the labor in such problems.

Down the left side of the table is a column headed "n". The "n" column is used for the number of interest periods.

Across the top of the table are listed four rates of interest. In the table the entry in the n = 40 row, $1\frac{1}{4}$% column is 1.64361946. That entry is used to determine the value of an investment after forty interest periods at a compound interest rate of $1\frac{1}{4}$%.

COMPOUND AMOUNT
When principal is 1

$$s = (1 + i)^n$$

n	$1\frac{1}{4}\%$	$1\frac{3}{8}\%$	$1\frac{1}{2}\%$	$1\frac{5}{8}\%$	n
1	1.0125 0000	1.0137 5000	1.0150 0000	1.0162 5000	1
2	1.0251 5625	1.0276 8906	1.0302 2500	1.0327 6406	2
3	1.0379 7070	1.0418 1979	1.0456 7838	1.0495 4648	3
4	1.0509 4534	1.0561 4481	1.0613 6355	1.0666 0161	4
5	1.0640 8215	1.0706 6680	1.0772 8400	1.0839 3388	5
6	1.0773 8318	1.0853 8847	1.0934 4326	1.1015 4781	6
7	1.0908 5047	1.1003 1256	1.1098 4491	1.1194 4796	7
8	1.1044 8610	1.1154 4186	1.1264 9259	1.1376 3899	8
9	1.1182 9218	1.1307 7918	1.1433 8998	1.1561 2563	9
10	1.1322 7083	1.1463 2740	1.1605 4083	1.1749 1267	10
11	1.1464 2422	1.1620 8940	1.1779 4894	1.1940 0500	11
12	1.1607 5452	1.1780 6813	1.1956 1817	1.2134 0758	12
13	1.1752 6395	1.1942 6656	1.2135 5244	1.2331 2545	13
14	1.1899 5475	1.2106 8773	1.2317 5573	1.2531 6374	14
15	1.2048 2918	1.2273 3469	1.2502 3207	1.2735 2765	15
16	1.2198 8955	1.2442 1054	1.2689 8555	1.2942 2248	16
17	1.2351 3817	1.2613 1843	1.2880 2033	1.3152 5359	17
18	1.2505 7739	1.2786 6156	1.3073 4064	1.3366 2646	18
19	1.2662 0961	1.2962 4316	1.3269 5075	1.3583 4664	19
20	1.2820 3723	1.3140 6650	1.3468 5501	1.3804 1977	20
21	1.2980 6270	1.3321 3492	1.3670 5783	1.4028 5160	21
22	1.3142 8848	1.3504 5177	1.3875 6370	1.4256 4793	22
23	1.3307 1709	1.3690 2048	1.4083 7715	1.4488 1471	23
24	1.3473 5105	1.3878 4451	1.4295 0281	1.4723 5795	24
25	1.3641 9294	1.4069 2738	1.4509 4535	1.4962 8377	25
26	1.3812 4535	1.4262 7263	1.4727 0953	1.5205 9838	26
27	1.3985 1092	1.4458 8388	1.4948 0018	1.5453 0810	27
28	1.4159 9230	1.4657 6478	1.5172 2218	1.5704 1936	28
29	1.4336 9221	1.4859 1905	1.5399 8051	1.5959 3868	29
30	1.4516 1336	1.5063 5043	1.5630 8022	1.6218 7268	30
31	1.4697 5853	1.5270 6275	1.5865 2642	1.6482 2811	31
32	1.4881 3051	1.5480 5986	1.6103 2432	1.6750 1182	32
33	1.5067 3214	1.5693 4569	1.6344 7918	1.7022 3076	33
34	1.5255 6629	1.5909 2419	1.6589 9637	1.7298 9201	34
35	1.5446 3587	1.6127 9940	1.6838 8132	1.7580 0275	35
36	1.5639 4382	1.6349 7539	1.7091 3954	1.7865 7030	36
37	1.5834 9312	1.6574 5630	1.7347 7663	1.8156 0207	37
38	1.6032 8678	1.6802 4633	1.7607 9828	1.8451 0560	38
39	1.6233 2787	1.7033 4971	1.7872 1025	1.8750 8857	39
40	1.6436 1946	1.7267 7077	1.8140 1841	1.9055 5875	40
41	1.6641 6471	1.7505 1387	1.8412 2868	1.9365 2408	41
42	1.6849 6677	1.7745 8343	1.8688 4712	1.9679 9260	42
43	1.7060 2885	1.7989 8396	1.8968 7982	1.9999 7248	43
44	1.7273 5421	1.8237 1999	1.9253 3302	2.0324 7203	44
45	1.7489 4614	1.8487 9614	1.9542 1301	2.0654 9970	45
46	1.7708 0797	1.8742 1708	1.9835 2621	2.0990 6407	46
47	1.7929 4306	1.8999 8757	2.0132 7910	2.1331 7387	47
48	1.8153 5485	1.9261 1240	2.0434 7829	2.1678 3794	48
49	1.8380 4679	1.9525 9644	2.0741 3046	2.2030 6531	49
50	1.8610 2237	1.9794 4464	2.1052 4242	2.2388 6512	50

Reprinted, by permission, from Stephen P. Shao, ed., *Mathematics for Management and Finance*, 2nd ed., South-Western Publishing Co., 1969, Appendix pp. 33–44. Adapted from Charles H. Gushee, *Financial Compound Interest and Annuity Tables*, Financial Publishing Company, Boston, 1958, and James W. Glover, *Compound Interest and Insurance Tables*, Wahr's University Bookstore, Ann Arbor, 1957.

Tables and Graphs   45

Returning to the investment of $10,000 mentioned earlier, the following computation would be used:

1. At an annual 5% interest rate, the rate figured quarterly is one-fourth of 5% of $1\frac{1}{4}$%.

   $(\frac{1}{4} \cdot 5\% = \frac{5}{4}\% = 1\frac{1}{4}\%)$

2. In ten years, interest figured quarterly means there are 10 times 4 or 40 interest periods.

3. The table entry for n = 40 at $1\frac{1}{4}$% is 1.64361946.

4. Multiplying the original investment $10,000 by the table entry 1.64361946 gives 16436.1946.

5. After ten years the original investment has a value of $16,436.19.

The $1\frac{1}{4}$% column of the table is appropriate for many transactions in our society today. Many department stores and credit card companies charge $1\frac{1}{2}$% per month on the unpaid balance. The dollar interest each month may seem reasonable, but the n = 12 at $1\frac{1}{2}$% entry is 1.19561817. The entry rounded off to two decimal places is 1.20 and means that $1\frac{1}{2}$% compounded per month is almost 20% annual interest.

Progress Test 7

1. Use the fraction-decimal table to find an approximation for $\frac{5}{13}$.

2. Use the table to write $\frac{2}{7}$ as a percent.

3. If a special kind of material costs $4.57 per yard, use the table of fraction-decimal approximations to find the cost of 28 inches (28 inches is $\frac{7}{9}$ yards).

4. Find the value of a $5,000 investment after 8 years at an annual rate of 6% compounded quarterly.

Exercise Set 4

I. 1. Use the fraction-decimal approximation chart to answer the following questions:

   a. What is the decimal approximation for $\frac{4}{11}$?

   b. Find the decimal approximation for $\frac{10}{15}$.

   c. Which has the greatest decimal approximation, $\frac{8}{13}$ or $\frac{11}{16}$?

   d. What fraction has a denominator of 14 and is approximately equivalent to .71?

   e. What fraction has a numerator of 9 and is approximately equal to .69?

   f. A farmer loses $\frac{1}{9}$ of his cattle in a snow storm. What percent of his cows were lost?

   g. In a class of 15 people, 13 were brunettes. What percent of the class were brunette?

   h. A car owner was able to increase the gas mileage on his car from 16 mpg to 19 mpg by changing the adjustments on the carburetor. By what percent did he increase the mileage?

   i. 10 out of a class of 14 people had decided on a major in college. What percent of the class has not decided on a major?

2. Find the value of a $10,000 investment after 10 years at an annual rate of 6% compounded quarterly.

3. If $4,000 is deposited in a savings account that pays 5% compounded quarterly, how much will be in the account at the end of 11 years?

4. What will be the value of an investment of $25,000 after 12 years at $6\frac{1}{2}$% compounded quarterly?

II. Challenge Problems

1. How much interest will be earned by an investment of $95,000 at $6\frac{1}{2}\%$ compounded quarterly for 5 years?

2. Give the decimal approximation for the difference between $\frac{11}{13}$ and $\frac{6}{16}$.

3. Sam made 5 hits out of 14 times at bat while George made 6 hits out of 17 times at bat. Which person had the best batting average? What was the difference in their averages?

4. A machinist was to cut a bolt 28 mm long but it was cut 30 mm long instead. What was the percent of error in the cut?

MODULE SELF-TEST

1. Use the table below to answer the following questions:

    a. What areas have experienced the greatest percent drops for March, 1975?

    b. What area has experienced the greatest percent drop in 1975?

    c. How many new dwelling units were produced in Dunedin in March, 1974?

Pinellas County Index of Economic Trends

New Dwelling Units	Mar. 1975	Mar. 1974	% Chg.	YTD 1975	% Chg.
St. Petersburg	73	317	— 77	217	— 60
County, rural	22	510	— 96	106	— 93
Clearwater	24	67	— 64	27	— 82
Largo	5	4	+ 25	11	— 98
Pinellas Park	18	—	*	25	— 84
Dunedin	3	67	— 96	12	— 85
Total	145	965	— 85	398	— 87

48   Introductory College Mathematics

2. Using the bar graph below, how many deaths per 100,000 residents were experienced in Tampa?

DEATHS FROM INFLUENZA AND PNEUMONIA PER 100,000 RESIDENTS

BEST
- San Jose — 21.31
- Tulsa — 23.01
- San Diego — 23.10
- Nashville — 25.23
- Honolulu — 27.70

WORST
- Atlanta — 62.32
- Buffalo — 62.45
- Tampa — 65.88
- St. Louis — 67.18
- Boston — 80.96

3. Using the graph below, what three trends were down in 1975 compared to 1974?

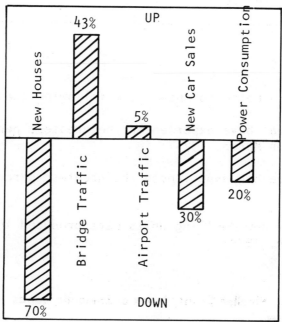

Trends: March, 1975 vs. March, 1974

4. Use the following graph to answer the questions:

a. Give the March, 1975 unemployment rates for: young adults, non-whites.

b. Give the 1974 unemployment rates for: adult men, teenagers.

Tables and Graphs 49

UNEMPLOYMENT RATES
Selected Categories

[ ] March, 1975
▨ 1974 avg.

	Household Heads	Adult Men	Adult Women	All Workers	Young Adults (20-24)	Nonwhite	Teen-agers (16-19)
March, 1975	5.8	6.8	8.5	8.7	14.3	14.2	20.6
1974 avg.	3.3	3.9	5.5	5.6	9.0	9.9	16.0

First National Bank of Boston and U.S. Bureau of Labor Statistics

5. Using the information from the graph below, if a man earned $3.00 per hour, what would be a good estimate of his job satisfaction rating.

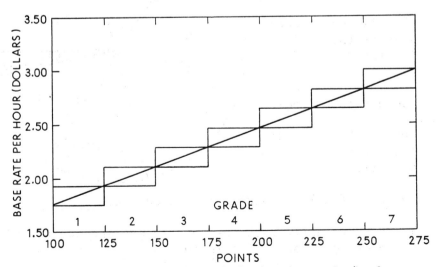

One arrangement of labor grades based on the regression line. Some companies use wider wage spreads for each grade so that the upper wage limit for one grade is above the lower limit of the next higher grade.

50 Introductory College Mathematics

6. The points A, B, C, and D in the graph below represent production for _____ profit.

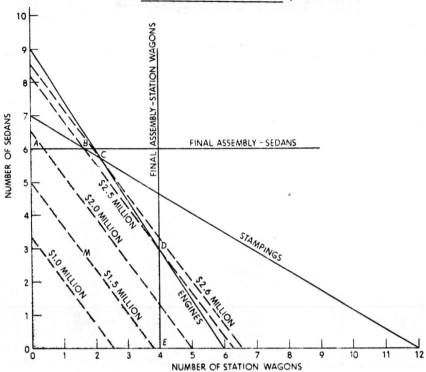

Chart showing possibilities of combinations for making sedans and station wagons.

7. The circle graph on the right shows the way a family with a monthly income of $1,200 spent its money.

   a. What is the correct size of the angle for food?

   b. How much did they spend for housing?

   c. What is the correct size of the angle for utilities?

   d. The sum of all the angle sizes should be _____ ?

   e. If another family used the same budget and spent $350 for housing, what is its income?

8. Use the computer below to find the miles per gallon if a 200 mile trip uses 12 gallons of fuel.

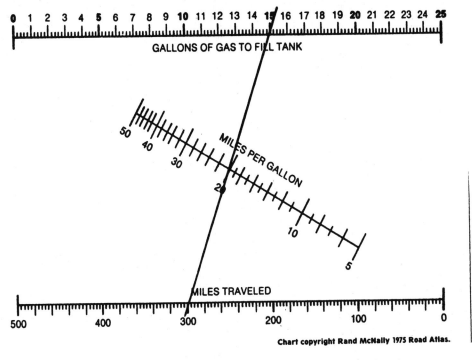

9. Use the fraction-decimal table to find the decimal equivalent to:

    a. $\frac{5}{11}$ b. $\frac{9}{13}$

10. Use the compound interest table to find the value of $1,000 savings left at $6\frac{1}{2}$% interest compounded quarterly for 9 years.

11. From the following graph, what is the relative amount of work done at:

    a. 100°F 30% relative humidity

    b. 100°F 100% humidity

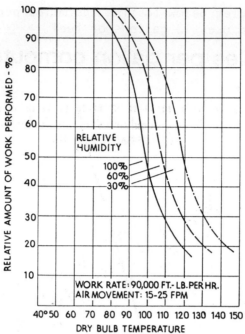

Temperature and humidity govern work a man can do. He works as well at 100°F., 30% relative humidity as at 82°F., 100% relative humidity.

12. Using the weather map below, is Colorado expected to receive average, above average, or below average precipitation the next 30 days?

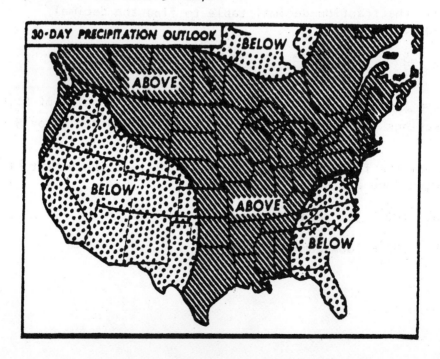

PROGRESS TEST ANSWERS

PROGRESS TEST 1

1. $99.95

2. 3

3. Model 12R

4. Adds, subtracts, multiplies, and divides, and has a floating decimal.

PROGRESS TEST 2

1. United States, Sweden, German Republic, Finland, Italy, and Mexico

2. 1971 - 1972

3.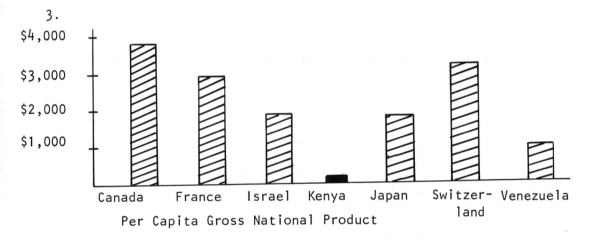

PROGRESS TEST 3

1. a. approximately 10
   b. 2

2. a. 1995 approximately
   b. 240 million

3.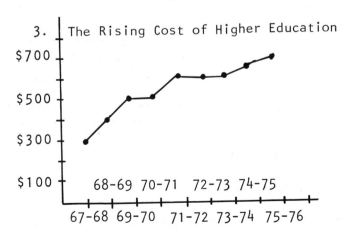

54 Introductory College Mathematics

PROGRESS TEST 4

1. a. 1 and 2   b. 1   c. 3

2.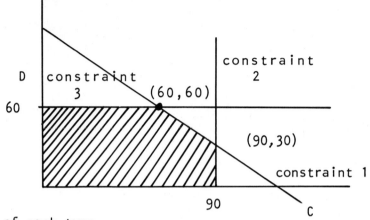

3. 60 of each type

PROGRESS TEST 5

1. yes

2. The arrows indicate the average of each group of data.

3. 14

4.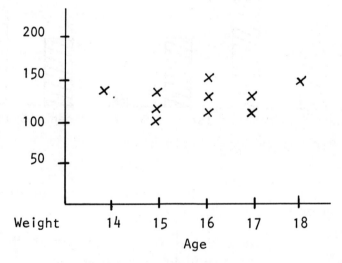

   a. zero   b. The weight tends to remain the same regardless of age.

PROGRESS TEST 6

1. a. 60%

   b. 40% more girls are mature than boys.

Tables and Graphs   55

PROGRESS TEST 6 (continued)

    c.   10% less boys are mature than girls

2.   a.   15   b. Unlimited branch banking

3.

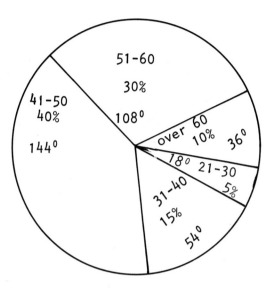

PROGRESS TEST 7

1.   0.38⁺   2. 29%   3. $3.56   4. $8,051.62

EXERCISE SET ANSWERS

EXERCISE SET 1

1.1.a. $460   3. a. $42.40

   b. $250,000   b. not listed

   c. $180   c. $5.82

2. a. 11.9%   d. $123.28

   b. 13.5%   4. a. 812

   c. 7.3%   b. up

   d. 21.1%   c. 6

   e. North and West
       10.0 and over

Introductory College Mathematics

EXERCISE SET 1 (continued)

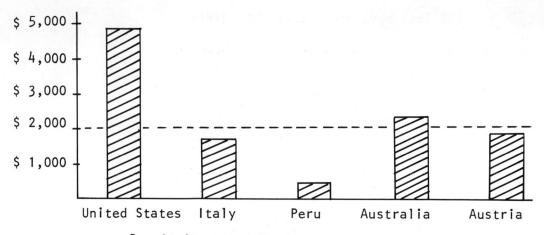

Per Capita Gross National Product 1970

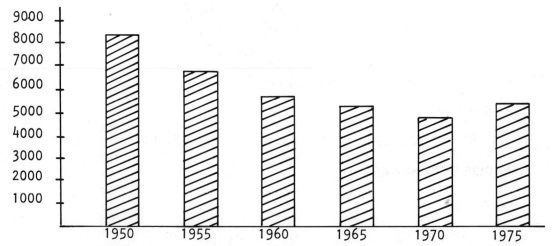

Number of People Employed in Agriculture (given in thousands.)

7. a. 3.5

   b. 4.2

   c. Men not on disability aged 45 - 64

   d. Seven times as many men die as women

11. CHALLENGE PROBLEMS

1. 8, 5

2. $3.93

CHALLENGE PROBLEMS (continued)

3. 4.3%

4. $247,450

5. $33\frac{1}{3}$

EXERCISE SET 2

1.1.a. It rose slightly and then returned to the same price.